GORILLAS

by Jaclyn Jaycox

PEBBLE
a capstone imprint

Pebble Explore is published by Pebble, an imprint of Capstone.
1710 Roe Crest Drive
North Mankato, Minnesota 56003
www.capstonepub.com

Copyright © 2020 by Capstone.
All rights reserved. No part of this publication may be reproduced in whole or in part, or stored in a retrieval system, or transmitted in any form or by any means, electronic, mechanical, photocopying, recording, or otherwise, without written permission of the publisher.

Library of Congress Cataloging-in-Publication data is available on the Library of Congress website.
ISBN: 978-1-9771-1342-9 (library binding)
ISBN: 978-1-9771-1795-3 (paperback)
ISBN: 978-1-9771-1350-4 (eBook PDF)

Summary: Simple text and photographs present gorillas, their body parts, and behavior.

Photo Credits
iStockphoto: hypergurl, 21; Shutterstock: Alan Tunnicliffe, 11, BDMPhoto, 15, Brina L. Bunt, 8, Dmitry Pichugin, 19, dptro, right 7, Edwin Butter, 20, Erni, 5, FCG, 10, GUDKOV ANDREY, 9, guentermanaus, 27, islavicek, 12, Joca de Jong, 24, Kiki Dohmeier, 16, Lisa Fitzthum Photography, 25, Marian Galovic, 1, Nick Fox, 22, Onyx9, Cover, left 7, Travel Stock, 17

Editorial Credits
Hank Musolf, editor; Dina Her, designer; Morgan Walters, media researcher; Tori Abraham, production specialist

All internet sites appearing in back matter were available and accurate when this book was sent to press.

Printed and bound in China.
002489

Table of Contents

Amazing Gorillas 4

Where in the World 6

Great Big Gorillas 10

On the Menu 16

Life of a Gorilla 20

Dangers 26

    Fast Facts 29

    Glossary 30

    Read More 31

    Internet Sites 31

    Index 32

Words in **bold** are in the glossary.

Amazing Gorillas

Imagine you could be any animal. What would you be? If you chose a gorilla, you are already close! They are a lot like us!

Gorillas are a type of **mammal**. Mammals can breathe air. Their body temperature is always the same. Females feed milk to their young.

Gorillas are the largest **primates** in the world. Primates are smart mammals. Humans are primates too!

Where in the World

Gorillas are found in Africa. There are two main types. Eastern gorillas live in east Africa. Western gorillas live in west Africa.

eastern gorilla

western gorilla

Eastern gorillas are bigger. Western gorillas are smaller. Eastern gorillas have long faces. Western gorillas have darker hair.

Most gorillas live in hot **tropical** areas. They live in rainforests. Others live in mountain forests. They are the only ones that live in cooler areas.

Gorillas mostly live on the ground. Smaller gorillas climb trees. They play in trees. They also get fruit from them. Young gorillas will sleep in trees. They make nests on the ground too. They sleep in their nests.

Great Big Gorillas

Gorillas are big. They are 4 to 6 feet (1.2 to 1.8 meters) tall. Males weigh almost 500 pounds (227 kilograms). Females weigh about half as much as males.

mountain gorilla

Gorillas have dark skin. They have brown or black hair. Mountain gorillas have longer hair. It keeps them warm. When males get older, the hair on their backs turns silver. These gorillas are called silverbacks.

Gorillas have long arms. They are longer than their legs! Their upper body is six times stronger than a human's.

Their hands are like human hands. They have four fingers and a thumb. This helps them climb trees and grab food. They are known for their "knuckle-walk." They use the backs of their fingers to walk.

Just like humans, gorillas have 32 teeth. They have a set of baby teeth first. Later they get the teeth they will keep for the rest of their lives. They have strong jaws. They chew lots of plants.

Instead of claws, gorillas have nails. They use them for scratching. They clean with their nails. They also scrape with nails. Gorillas use them to pull out old hair. They scratch away grass.

Gorillas have big heads and small ears. Every gorilla has a different nose. No nose has the same wrinkles.

On the Menu

Gorillas need a lot of food. They can eat more than 40 pounds (18 kg) of food in a day. Male gorillas eat more than females.

They eat different kinds of plants. Gorillas eat bark from trees. They eat leaves too. Sometimes they eat ants and termites. They also eat many different fruits. They are known to eat more than 100 different kinds!

Gorillas that live in different places eat differently. Gorillas that live in mountains eat more plants. Gorillas that live in lower places eat more fruit.

They rarely drink water. Many of the plants they eat have water in them.

Life of a Gorilla

Gorillas live in groups called troops. A troop usually includes a silverback, a few younger males, and females and their young. A troop can have two to 30 members. The silverback is the leader of the group.

Gorillas are peaceful animals. But a silverback will scare away other gorillas if he feels **threatened**. He will beat his chest and scream. This makes others run away.

Silverbacks **mate** with the females. Females usually give birth to one baby. Baby gorillas are called infants. They weigh about 4 pounds (1.8 kg). They drink milk from their mothers until they are about 3 years old.

Babies grow fast. They can usually walk by eight months old. But they ride on their mom's back until they are 2 or 3 years old.

Young gorillas stay close to their mothers. They are never more than a few steps apart. Mom and baby share a nest for up to six years.

Young gorillas learn by watching the others in the troop. They learn how to get food. They learn how to care for baby gorillas. They also love to play. They stay with their moms until they are 7 to 10 years old. Then they leave to join or start another troop. They can live up to 35 years in the wild.

Dangers

Gorillas don't have many **predators**. Crocodiles and leopards may hunt them. But humans are their greatest danger. People hunt them for food. Forests are being cut down. Their homes are being destroyed. Sicknesses can also kill them.

Humans cut down forests where gorillas live.

Gorillas are in danger of dying out. They are **endangered**. But people are working to help them. They protect the forests where they live. Laws are being passed to keep them from being hunted. Some people help sick or hurt gorillas. They help them get healthy. They help return them to the wild.

Fast Facts

Name: gorilla

Habitat: tropical rainforests, mountain forests

Where in the World: Africa

Food: plants, ants, termites, fruit

Predators: leopards, crocodiles, humans

Life span: 35 years

Glossary

endangered (in-DAYN-juhrd)—in danger of dying out

mammal (MAM-uhl)—a warm–blooded animal that breathes air; mammals have hair or fur; female mammals feed milk to their young

mate (MATE)—to join with another to produce young

predator (PRED-uh-tur)—an animal that hunts other animals for food

primate (PRYE-mate)—any member of the group of intelligent animals that includes humans, apes, and monkeys

threaten (THRET-uhn)—to do something that causes another to feel in danger

tropical (TRAH-pi-kuhl)—hot and wet places; places near the equator are tropical

Read More

Bové, Jennifer. *I Wish I Was a Gorilla*. New York: HarperCollins Publishers, 2018.

Hall, Margaret. *Gorillas and Their Infants: A 4D Book*. North Mankato, MN: Capstone Press, 2018.

Marsico, Katie. *Gorillas*. New York: Scholastic, Inc., 2017.

Internet Sites

Cool Kid Factz—Gorilla Facts
coolkidfacts.com/gorilla-facts/

National Geographic Kids—Mountain Gorilla
kids.nationalgeographic.com/animals/mountain-gorilla

San Diego Zoo Kids—Western Lowland Gorilla
kids.sandiegozoo.org/index.php/animals/western-lowland-gorilla

Index

Africa, 6

bodies, 13

Eastern gorillas, 6, 7
eating, 16, 17, 18

forests, 8, 26, 28
fruit, 9, 17, 18

hair, 7, 11, 14

mating, 23
milk, 4, 23
mountains, 8, 11, 18

nails, 14
noses, 15

plants, 14, 17, 18

silverbacks, 11, 20, 21, 23
size, 4, 7, 10

teeth, 14
threats, 21, 26, 28
troops, 20, 25

Western gorillas, 6, 7

young gorillas, 4, 9, 20, 23, 24, 25